Butterflies of West Texas Parks and Preserves

Butterflies of West Texas Parks and Preserves

Roland H. Wauer

Texas Tech University Press

This book was set in Goudy Oldstyle BT and Freehand 591 BT. The paper used in this book meets the minimum requirements of ANSI/NISO Z39.48-1992 (R1997). ♾

Printed in China

Design by Brandi Price

All photographs by the author

Library of Congress Cataloging-in-Publication Data

Wauer, Roland H.
Butterflies of West Texas parks and preserves / Roland H. Wauer.
p. cm.
Includes bibliographical references (p.).
ISBN 0-89672-471-9 (cloth : alk. paper) — ISBN 0-89672-472-7 (pbk. : alk. paper)
1. Butterflies—Texas, West—Identification. I. Title.
QL551.T4 W38 2002
595.78'9'097649—dc21

2001005021

02 03 04 05 06 07 08 09 10 / 9 8 7 6 5 4 3 2 1

Texas Tech University Press
Box 41037
Lubbock, Texas 79409-1037 USA

1-800-832-4042
ttup@ttu.edu
www.ttup.ttu.edu

I dedicate this book to Roy Kendall, an avid lepidopterist who dedicated so much of his life to the understanding of the butterflies of the Big Bend Country.

Contents

Preface

Learning to identify butterflies is not always an easy task. It sometimes takes time and patience, as well as a few tools, such as binoculars and a field guide. Beginners cannot help being frustrated when trying to identify a lone skipper, for instance, while using a field guide that includes a wide assortment of options. A dozen or more species, at least at first glance, are likely to be reasonably similar—thus the reason for the approach used in this field guide.

Butterflies of West Texas Parks and Preserves focuses on the most common species in the area covered and also provides suggestions about other likely candidates. For example, fourteen species of hairstreaks are found in Big Bend National Park, but just five of those can be expected with any regularity. By emphasizing those five species, this field guide makes identification more straight-forward than do other popular North American field guides, which might include more than six dozen options!

The intention of this field guide is to make butterfly identification in the West Texas region easier and, therefore, to encourage greater enthusiasm for an activity that is not only a healthy one but that also helps the participant appreciate the natural world. With appreciation comes understanding, and with understanding comes the desire to protect our parks and preserves.

Acknowledgments

Several individuals have provided assistance in helping the author learn and appreciate butterflies in recent years. I especially want to thank Paul Miliotis, Mike Overton, and Bob Pyle.

The manuscript for this book has been reviewed by Mike Overton and Phil Schappert; their most useful suggestions are gratefully acknowledged. The map of West Texas parks and preserves was prepared by Mark Elwanger. His help is much appreciated.

Introduction

Butterflies are those flying gems that we see almost everywhere we go in the outdoors. They occur year-round in the varied habitats of West Texas and are some of nature's most exquisite creations. They come in a vast assortment of colors: bright yellows, blues, and oranges to dull grays and browns. And they can be as small as a dime or as large as a dollar bill.

Nearly two hundred butterfly species have been recorded within the West Texas region. Although many of these occur year round, others are present only at certain seasons or are casual strays whose occurrence cannot be expected. Fifty of the most common West Texas species are featured in this book, with a few additional species highlighted because they are unique to the region. These "specialties" are the ones most sought by butterfly enthusiasts.

All of the butterflies that have been recorded in the region are listed on the "Butterfly Checklist—West Texas Parks and Preserves" included in this book. This list can be used to learn which species may be seen in this region, how abundant they are, and where they are most likely to be found. It also can be used as a personal checklist of your findings.

Common and scientific butterfly names used in this book are those suggested by the North American Butterfly Association (NABA) (2001). Additional common names may also be included, especially when they are traditional names that are used by other authors in earlier well-known field guides. Plant names used follow *Checklist of the Vascular Plants of Texas* by Stephan L. Hatch and colleagues (1990).

Butterfly Basics

Almost eight hundred species of butterflies have been recorded in North America north of the Mexican border, and almost eleven thousand species of moths have been reported within this same area, a greater than 13 to 1 ratio. Both are members of the insect order Lepidoptera, a term meaning "scale wings." These scales give butterflies their marvelous colors and reflect sunlight at various angles. They readily come off the butterfly when handled and as it ages. Be aware that older individuals that have lost scales or a section of their wings due to accidents or bird strikes can look very different from a freshly emerged, bright, fully intact individual.

Both butterflies and moths possess three body sections—head, thorax, and abdomen—and both undergo complete metamorphosis. This includes four stages: (1) eggs that hatch into (2) larvae or caterpillars that feed on foodplants, and eventually become a (3) pupa or chrysalis, the resting stage in which a caterpillar transforms into an (4) adult butterfly or moth. Within a few hours after emergence, after expanding its wings, the adult, brighter and more perfect than it will be at any other time in its life, is airborne.

Butterflies differ from moths in a number of ways. Principally, butterflies are daytime fliers that possess thin antennae with a thickened clublike tip. Moths are usually nighttime fliers, are hairier than butterflies, and possess either feathery or filamentous antennae that taper at the end. Their flight is also quite different. Butterfly flight is usually fluttery and graceful while that of moths is often jerky or direct. Most butterflies land in the open on a flower, leaf, or rock, or on the ground, while moths generally land where they can readily hide, such as in a clump of grass or under a leaf. Most moths also seem to fall from flight rather than glide into a landing as butterflies do.

Adult butterflies usually live only a few days or weeks, although the complete generation—from egg to adult—may last a full year or more. Mourning Cloaks are said to be one of the longest-lived North American butterflies, able to survive in the adult stage for as long as eleven months, and some northern and alpine species may take two years for a complete generation.

The size of any one species can vary as much as 50 percent, depending upon the nutrients available to the caterpillar. Also, since butterfly colors often begin to fade in only a few days, a butterfly's pattern, rather than its colors, may be most important for identification purposes.

During their relatively short lifetimes, most butterflies obtain nourishment from a variety of sources, although a few species never take food as adults. Caterpillars feed by chewing vegetation, and adults feed by drinking their nutrients and moisture. Most take nectar from flowers, while others find nutrients in rotting fruit or animal materials, such as a carcass or scat. Some species also obtain nutrients from tree sap, and their moisture and salts from mud.

The most important task for an adult butterfly is to find a member of the opposite sex and mate. Some male butterflies, such as some longwings, may locate an emerging female and mate even before she has fully emerged from the chrysalis. After mating, the female must find the appropriate foodplant(s) on which to lay eggs so that the resultant caterpillars can obtain a suitable food supply.

Butterflies possess excellent eyesight for detecting movement and are able to see an approaching predator or butterfly enthusiast quite well. Butterflies also see the ultraviolet range of colors. Although they are not able to detect odors as humans can, they utilize their antennae to "smell," and they "taste" with their feet. Once a source of nutrients is located, they drink the nectar or other materials with their long, tonguelike proboscis.

Butterflies face a variety of threats throughout their lifetime. Eggs and caterpillars are taken by a multitude of vertebrate and invertebrate predators, and over 50 percent of some broods may be parasitized. Hazards to adult butterflies exist almost everywhere. While in flight, they can be captured by a bird, robber fly, or dragonfly, and when perched on a flower or other substrate, they are preyed on by spiders and ambush bugs that may be hidden in flower heads or waiting in a nearby web. Other wide-ranging predators include praying mantids, lizards, small rodents, and deadly house cats. There are also the abundant human-induced threats, ranging from the overzealous use of pesticides to the elimination of native habitats for a wide variety of human-made developments or habitat "improvements."

With the abundant dangers butterflies must face daily, from egg through the adult stage, it is a wonder that they are as numerous as they are.

Butterfly Watching

Butterfly watching is one of the fastest-growing hobbies in North America. But how does one get started? Although one can often enjoy butterflies without using any equipment by inching close enough to see them well, two things are essential to enjoy them more. First is the ability to observe them at a distance so that a close approach does not frighten them away. A close-focusing (within four to five feet) binocular, now available in a range of prices, is essential. High-quality binoculars are easier to use, cause less eyestrain, and are likely to last longer.

Second and equally important is a good field guide. Part of the enjoyment of butterflies, at least for many people, is the ability to put a name on each butterfly found. This book is designed to help identify West Texas species only, but eventually you will probably also want to obtain one or several field guides that include all possible species. These come in two styles: those that use paintings or photographed specimens, which usually display only the dorsal (upper) views; and those that utilize photos of live butterflies in their natural state. Both styles are useful, but the latter is most helpful in the field. This is especially true for butterflies that rest with folded wings, making them extremely difficult to identify because the ventral (underparts) view must be compared with a dorsal (upperparts) illustration. Several of the better reference books are listed in the "Other Field Guides" section at the end of this book.

In addition, butterfly photography is also becoming more popular. Photography not only allows you to enjoy your observations at a later time and to share them with friends, but also may help to identify or to document those species that are not so easily identified in the field or are out of range and need further verification. Butterfly photography is a broad topic that needs more discussion than can be given here. Personally, I utilize a Canon camera with a 300 mm lens that allows me to fill a frame with a dime-sized butterfly at about four feet. The best resolution is obtained with slow (50–100 ASA) film, so flash is essential in many cases.

When and Where

Much of finding and watching butterflies depends upon the season and weather, which control the availability of flowering plants. Finding nectar sources is often the first step in locating butterflies. However, the peak flowering season does not necessarily coincide with peak butterfly abundance. Wildflowers are most abundant in West Texas in spring but are more prevalent at higher elevations in summer and early fall. If the region has not received adequate rainfall, flowering plants can be few and far between at any time of year.

There is rarely a time when butterflies do not occur within the lowlands of West Texas. Although they do not fly during extremely cold periods in winter, as soon as temperatures reach into the high 50s or 60s, a few of the most hardy species, such as Checkered White, Sleepy Orange, Dainty Sulphur, and Variegated Fritillary, become active. A few species hibernate as adults in winter, while most others remain in the egg, larvae, or chrysalis stage.

The secret to finding butterflies in winter is to find a nectar source. A few flowering plants, such as cowpen daisy and globe mallow, can usually be found along the Rio Grande floodplain. At slightly higher elevations, open arroyos often support poreleaf shrubs. This latter species may be the single most important nectaring plant during December and January.

As the days become warmer, generally by mid-February, lowland flats, arroyos, and roadsides come to life, usually producing a variety of desert wildflowers. Immediately, a number of butterflies put in their appearance: Pipevine Swallowtail, Orange Sulphur, Southern Dogface, Sleepy Orange, Dainty Sulphur, Gray Hairstreak, Reakirt's Blue, Painted Lady, Red Admiral, Common Buckeye, Queen, Common Checkered-Skipper, Sachem, and Eufala Skipper are some of the earlier ones.

Gradually, as the length of days increases and temperatures grow warmer, flowers appear at higher elevations, and with them an increase in the number of butterflies. By March and early April, except in very cold or very dry years, as many as thirty-five to forty species usually can be found in a single day. However, few species are found in the higher, cooler canyons until late April or May. Additional species to be expected in spring include Lyside Sulphur; Mexican Yellow; Western Pygmy-Blue; Marine Blue; Fatal Metalmark; American Snout; Tiny and Elada Checkerspots; Texan Crescent; Red-spotted Admiral; California Sister; Tropical Leafwing; Empress Leilia; Golden Banded-Skipper; Northern Cloudywing; Mournful Duskywing; Acacia, Clouded, Fiery, and Sheep Skippers; Common and Scarce Streaky-Skipper; and Bronze Roadside-Skipper.

Late spring to midsummer (mid-April through August) usually produces the greatest variety of mountain species, especially when late winter and/or spring rains have occurred. If rains do not occur, butterfly numbers will be low, and those individuals will usually be found only in moist areas such as at springs or in the mountain canyons. Two-tailed Swallowtail, Mexican Yellow, Red-spotted Admiral, California Sister, Tropical Leafwing, Empress Leilia, and Red Satyr are the most obvious of the mountain species.

So long as the traditional summer rains occur, the fall months typically produce the most diverse butterfly fauna. Although these rains are often extensive, during dry years with scattered showers only, the butterfly watcher will have to seek out the best flowering areas.

A number of butterflies are present in late summer and fall that may not occur or are rarely present at other times. Examples include Mimosa Yellow, Leda Ministreak, Chinati Checkerspot, Common Mestra, Mead's Wood-Nymph, and Monarch. A little later in fall, there is a noticeable migration of a few species such as Southern Dogface, Large Orange Sulphur, American Snout, Queen, and Monarch.

Finding and Identifying Butterflies

Finding butterflies is a lot like finding birds. To locate the highest numbers requires visits to a variety of habitats. Although some species are widespread in their occurrence, and many wander widely, most are limited to localities where their foodplants exist. Therefore, the best chance for finding some of the less abundant species is to learn and seek out their essential foodplants.

Once a specific area is selected, walk slowly through it, watching ahead for butterflies. Use your binoculars to search for those that might be perched ahead in the open. Emperors, buckeyes, and various other species often rest in open areas. Approach them slowly to within eight or ten feet for a better look through your binoculars. Be careful that your shadow does not fall across a resting butterfly. Stop at each flowering plant along the way to see what smaller species might be nectaring there. Keep in mind that some species blend into their environment remarkably well and may not be easily detected. A single flowering shrub can hold a dozen or more individuals.

At times even a careful approach will flush a butterfly that you especially wanted to see well. That individual will often return to the same plant if you stop and wait. Be patient. You may also be able to follow its flight to another perch. This is sometimes difficult but becomes easier with practice.

In addition, it helps to understand butterfly behavior. Although this topic is an extremely broad one, there are a few basic tips that can be used in finding various species. Many butterflies patrol certain areas in search of a mate. Empress Leilias, California Sisters, and buckeyes, for instance, can often be found flying up and down arroyos. Others, such as Red-spotted Admirals, Red Admirals, Golden Banded-Skippers and Umber Skippers, and various duskywings, perch on rocks and/or shrubs and trees, and wait for a passing mate.

Some butterflies, especially giant-skippers, "hilltop," meaning that they fly to elevated landmarks with an extensive view to wait for a passing mate. Some may remain there all day, but others retire to more shaded sites once temperatures become warmer.

Temperatures do affect butterfly activity; most are less active during the heat of the day than during midmorning or late afternoon. Some species may spend their morning hours out in the open, either on a hilltop or at the top of a tree, but retire to a shaded area at midday.

Again, knowledge of a butterfly's adult food and foodplants can be extremely important in locating individual species. Although most species are generally found at various nectar sources or as flybys, knowing the larval foodplants and where they occur may be essential.

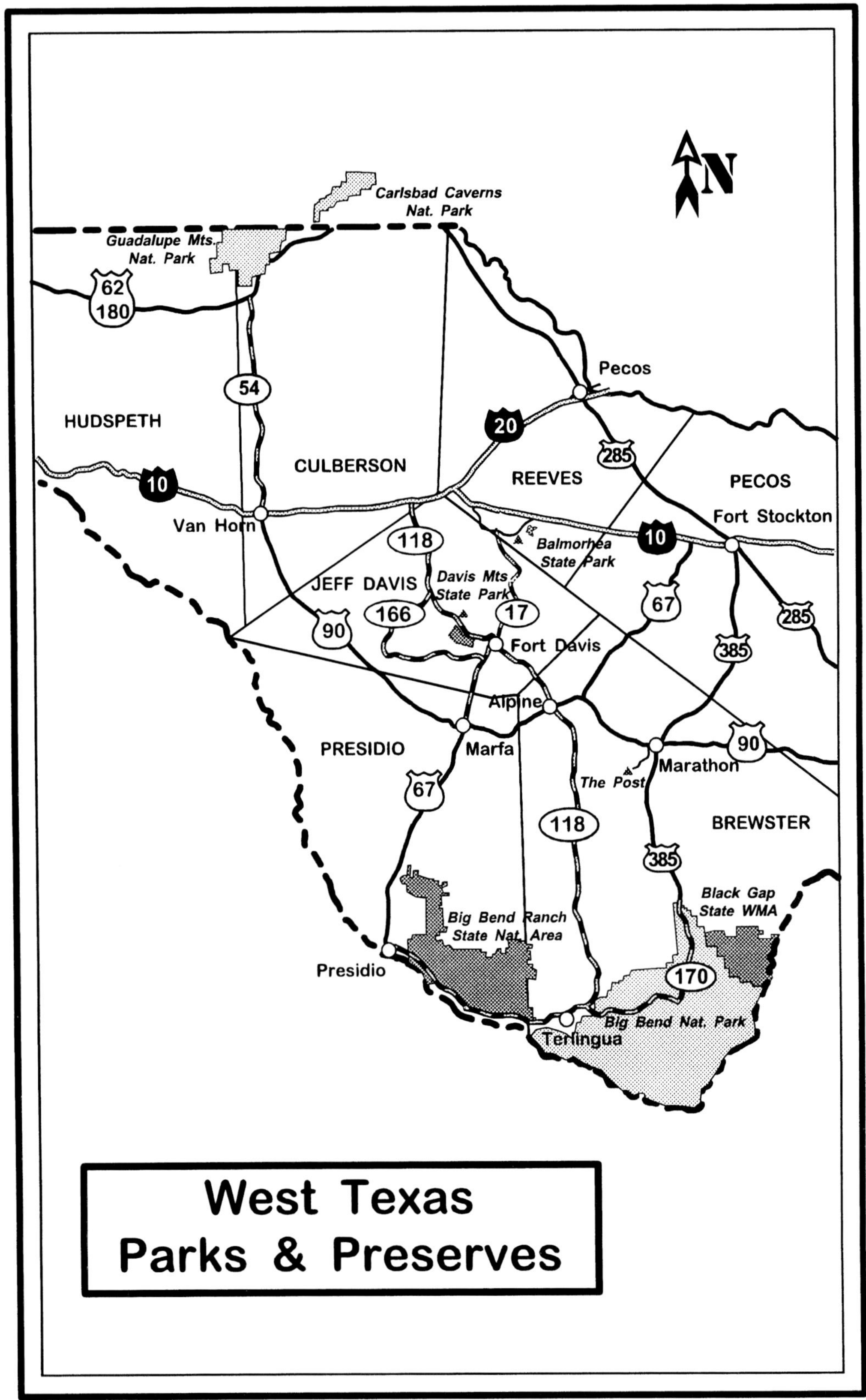

N
Carlsbad Caverns Nat. Park
Guadalupe Mts. Nat. Park
62
180
54
HUDSPETH
CULBERSON
Pecos
20
285
REEVES
PECOS
10
Van Horn
Fort Stockton
10
118
Balmorhea State Park
JEFF DAVIS
Davis Mts State Park
67
166
17
90
Fort Davis
285
385
Alpine
PRESIDIO
Marfa
90
Marathon
The Post
67
118
BREWSTER
385
Black Gap State WMA
Big Bend Ranch State Nat. Area
Presidio
170
Big Bend Nat. Park
Terlingua
West Texas
Parks & Preserves

Butterflies of West Texas Parks and Preserves

The following fifty butterfly species are those you are most likely to see throughout the West Texas region, in both the desert lowlands and in the adjacent mountains. Look-alike species are also mentioned in the discussions so that you can use a more general field guide (see suggestions in "Other Field Guides" at the end of this book) to help identify similar but less commonly encountered individuals. Also, a comprehensive checklist of the West Texas butterflies is included in the last section of this book.

Swallowtails

FAMILY PAPILIONIDAE

Members of the Swallowtail family in Texas are always large and showy, have six visible walking legs, and often possess one or two tails on each hindwing. Adults fly relatively slowly, generally five to seven feet above the ground. While feeding or nectaring, they constantly flutter their wings, and they (especially males) frequently sip water and nutrients at mud puddles.

Pipevine Swallowtail *Battus philenor*

Wingspan, 2.75–4 inches. This is the large black-and-blue butterfly also known as Blue Swallowtail in some books. It possesses a single, all-dark tail on each hindwing. In flight, it shows a glossy, metallic blue-green color on the upper hindwings. The underparts possess a single band of orange spots against a metallic blue-black background on the hindwings and smaller whitish spots against a blackish background on the forewings.

SIMILAR SPECIES: There are two possibilities: Black Swallowtail and Red-spotted Admiral. Black Swallowtails lack the broad metallic-blue hindwings above and show two rows of yellow dots above and two rows of red-and-yellow dots below. Red-spotted Admirals normally occur only in the mountains and are tailless. They often perch out in the open on foliage, with folded wings showing red marginal and basal spots (see further description below).

WHEN AND WHERE: Pipevine Swallowtails can appear almost anywhere, but they are most often found flying in the lowlands in spring and late fall and at higher elevations in summer. Larval foodplants are limited to pipevines *(Aristolochia)* that possess aristolochic acids, which are ingested by the caterpillars. Pipevine Swallowtail larvae and adults, therefore, are distasteful to predators. Two species of pipevines are found in West Texas: Cory and Wright's Dutchman's pipes both occur in rocky areas in the mountains.

REMARKS: The flight of these large butterflies can be fairly rapid, with shallow wing beats. Males patrol for females near foodplants or on hilltops.

Two-tailed Swallowtail *Papilio multicaudatus*

Wingspan, 3.5–4.25 inches. This highland swallowtail possesses bright yellow upper wings with narrow black tiger stripes. Both forewings and hindwings have black margins with a yellow submarginal band, and hindwings have blue patches. This butterfly has a double tail; the outer tail is much longer than the inner.

SIMILAR SPECIES: Two-tailed Swallowtails can be confused in West Texas with only the very rare (in our area) Eastern Tiger Swallowtail, which possesses just one tail. Tigers also show broader (compared with Two-tailed) black stripes on their forewings. Two other yellow-and-black swallowtails do occur: Giant and Ornythion. Both are brownish black with a broad diagonal band of yellow blotches from wing tip to wing tip. The Giant Swallowtail possesses a dark tail with a yellow center, while the Ornythion Swallowtail has an all-dark tail.

WHEN AND WHERE: Two-tailed Swallowtails are primarily a warm-weather species that occurs almost exclusively in the highlands, especially in wooded canyons. They are occasionally found in the lower canyons and even more rarely at mid-elevation springs and along the Rio Grande. Larval foodplants include a number of species of the rose and citrus families, such as serviceberry, mountain mahogany, Apache plume, chokecherry, and various wild roses, all of which are mountain species.

REMARKS: This lovely creature patrols canyon bottoms and woodland openings during the daylight hours. Its floating flight is often fairly high above the terrain. However, after summer rains, several individuals may perch at wet spots in canyon bottoms, obtaining moisture and nutrients.

Whites and Sulphurs

FAMILY PIERIDAE

Members of this family are small to medium sized, lack fully developed tails, and possess three visible pairs of walking legs with forked claws. Their flight is steady and often in a straight line.

Whites—Subfamily Pierinae

Whites are predominantly white in color, although all members possess black or dark spots or smudges. Some individuals also possess some yellow. Whites utilize plants of the Mustard family for larval foodplants.

Checkered White *Pontia protodice*

Wingspan, 1.5–2 inches. One of the few mostly white butterflies possible, Checkered Whites possess rather large black spots on the outer half of their forewings. Females are more yellowish than the males.

SIMILAR SPECIES: Four other Whites are similar—Mexican Dartwhite, Florida White, Cabbage White, and Great Southern White—but all these others are extremely rare and should not be expected.

WHEN AND WHERE: This may be the most numerous wintertime butterfly in the lowlands; it remains fairly common into spring and is often common again in late fall. By March and April it begins to appear at higher elevations. Larval foodplants include a wide assortment of the Brassicaceae (Cruciferae) family, mainly desert species, including mustards, bittercress, drabas, wallflowers, pepperweeds, bladderpods, prince's plume, twistflowers, and Gregg keelpod.

REMARKS: The flight of Checkered Whites is typically swift and erratic and usually fairly low to the ground. Occasionally they will alight to take nectar, and that provides the best opportunity to study them well.

Sulphurs—Subfamily Coliadinae

Orange Sulphur *Colias eurytheme*

Wingspan, 1.75–2.25 inches. This yellow to orange-and-black butterfly is also known as Alfalfa Butterfly. Females can be whitish green. A ventral view shows large and small rust-colored circles with silver centers on the hindwing, and a black circle or triangle with a silvery center on the forewing. Males possess a series of faint submarginal spots on both wings. A dorsal view reveals broad black margins with orangish spots on the female's wings, and non-spotted black margins on the male's wings. Both sexes have one central black dot on the forewings. The orange-yellow background color may be almost white in females.

SIMILAR SPECIES: Several of the medium-sized sulphurs can be confused with this species. Southern Dogface (described on the next page) and Cloudless Sulphur look similar, as does the rare Clouded Sulphur. The Cloudless Sulphur is all yellow with tiny rust-colored spots and usually flies only in summer. The Clouded Sulphur is slightly smaller and has two double rings with silver interiors on its ventral hindwings, and reasonably large brownish spots located submarginally on both wings. A dorsal view is very similar, but the broad dark edges of the Clouded Sulphur are brownish rather than black, and the background color is more yellow than orange.

WHEN AND WHERE: The Orange Sulphur occurs in small numbers year-round in the lowlands but is present at higher elevations only in summer. It is one of the more common wintertime butterflies, nectaring on flowering poreleaf plants in lowland arroyos. Larval foodplants include a wide assortment of legumes, especially herbaceous species such as daleas, locoweeds, lupines, tickclovers, and vetches.

REMARKS: This species occasionally hybridizes with Clouded Sulphurs, producing offspring that can make identification confusing.

Southern Dogface *Colias cesonia*

Wingspan, 2.1–2.6 inches. Sometimes known as Dogface Butterfly, it is easily identified when it can be viewed either with its wings spread or from the side; backlit individuals show a perfect silhouette of a poodle's face, complete with a black eye. The more typical dorsal view reveals a green-yellow butterfly with a slightly falcate (hooked) wing tip. Its ventral features are not unlike those of the Orange Sulphur—a central pinkish spot with a silver center on the hindwing and a larger black circle with a silver center on the forewing. Females show obvious pink streaks across the forewings, and females are rarely white.

SIMILAR SPECIES: Although this species is very distinct when seen well, it can possibly be confused with the Cloudless Sulphur and, less likely, with Orange and Clouded Sulphurs. See comments above.

WHEN AND WHERE: This is the most abundant of all the sulphurs in our area and can be expected year-round. It is most numerous in the lowlands but can be reasonably common in the mountains during the summer months as well. Larval foodplants include a wide assortment of legumes, but unlike the Orange Sulphur, the Dogface utilizes many of the woody species, including leadplant, kidneywood, and daleas.

REMARKS: Males patrol open areas in search for females. In summer, the Dogface often perches at puddles for moisture.

Lyside Sulphur *Kricogonia lyside*

Wingspan, 1.5–1.75 inches. This may be the most abundant of the region's spring, summer, and fall butterflies, ranging from the Rio Grande floodplain into the upper mountain canyons. It can be extremely variable in color, from a dull green and yellow to almost white. A golden wing base is often evident, especially in flight. The underside of the hindwing usually has a satiny sheen and noticeably thickened veins.

SIMILAR SPECIES: No other West Texas butterfly looks like a Lyside. Sulphurs of the same size and smaller, such as Boisduval's, Mexican, Little, and Mimosa yellows, are all well marked.

WHEN AND WHERE: This sulphur prefers heavy vegetation and is seldom found in the open. It seems to utilize thickets wherever they occur, such as the riparian growth along the Rio Grande, or along desert arroyos or in mountain canyons. Guayacan, a desert shrub that occurs up to an elevation of about forty-five hundred feet, is its principal larval foodplant.

REMARKS: Lyside is a tropical species that normally resides in the United States only in Texas, southern New Mexico, and southern Arizona. Because of the importance of guayacan as a larval food source, this butterfly is also known as Guayacan Sulphur.

Mexican Yellow *Eurema mexicana*

Wingspan, 1.5–2 inches. This is a pale yellow to creamy white butterfly with a rather distinct triangular shape, a tailed hindwing, and a rusty dusting on the underside of the hindwing. The dorsal side shows a heavy black pattern resembling a poodle face without the eye.

SIMILAR SPECIES: Boisduval's Yellow is only slightly smaller with a less pronounced tail projection. It is brighter yellow with more rust ventrally, and the dorsal side is lemon yellow with heavy black only along the borders. The very rare Tailed Orange is orange with black veins.

WHEN AND WHERE: Mexican Yellows occur year-round and at all elevations. They are absent from the highlands only in winter when they are most likely to be found in lowland arroyos, nectaring on poreleaf. Larval foodplants include acacias, most notably acacia, New Mexico locust, and senna.

REMARKS: This is a very active species that usually flies throughout the daylight hours, yet its flight is slower than that of most of the other small sulphurs.

The Mexican Yellow is known for its long-range dispersal northward in summer. In spite of its southern affinity, a few individuals are annually found as far north as Canada. It was once known as "wolf-face sulphur," due to the dog-face pattern on the wing.

Sleepy Orange *Eurema nicippe*

Wingspan, 1.5–1.85 inches. Sleepy Orange is one of the most numerous and widespread butterflies found in West Texas. When perched, it is best identified by its elongated shape, with just a hint of a tail projection, and cocoa brown to yellow-brown color with diagonal rust markings on the underside of the hindwing. The dorsal side shows a bright orange color with black borders.

SIMILAR SPECIES: The Tailed Orange, very rare in West Texas, holds itself much like the Sleepy Orange when perched but shows a noticeable tail projection during the winter months; in summer its tail is little more than a sharp-angled edge of the hindwing. Also, the Orange Sulphur can be confused with this species but has a very different outline.

WHEN AND WHERE: The Sleepy Orange can be found almost anywhere during most of the year; in winter it is most likely to be found in lowland arroyos where it nectars on poreleaf. Larval foodplants include a variety of legumes, although sennas are most often utilized.

REMARKS: In spite of its name, this is a fast-flying butterfly that is most often seen as a flyby, but even then can usually be identified by its distinct orange color.

Dainty Sulphur *Nathalis iole*

Wingspan, 0.75–1.15 inches. In spite of its small size, this species is difficult to miss most times of the year because it is usually so abundant. Its two-toned pattern is often evident even at a distance. In winter the Dainty Sulphur can be quite blackish green below; but in summer it is yellowish with a blackish blush on the hindwings and golden yellow forewings, with one large and several smaller black spots on the outer edge. Dorsally, it is yellow with broad black wing tips and a black bar at the trailing edge of the forewing and the leading edge of the hindwing.

SIMILAR SPECIES: Two other tiny yellowish butterflies occur in summer: Little and Mimosa Yellows. Both are rare. The slightly larger Little Yellow does not show a two-toned pattern any time of year; it is marked ventrally by a distinct red spot on the outer margin and two black spots at the base of the hindwings. A dorsal view shows a tiny black center spot and broad black edges on both wings. Mimosa Yellow is pale yellow and lacks the two black spots at the base of the hindwings and black borders on the hindwings.

WHEN AND WHERE: The Dainty Sulphur can be found almost anywhere, but it is most common in open areas, such as on flats, in broad arroyos, and along roads and trails. Larval foodplants include a variety of composites, including such common species as huisache-daisy and other daisies, dogweed, ragweed, brickellbush, rabbitbrush, and thistles.

REMARKS: Considered the smallest of all the sulphurs, this little butterfly typically flies close to the ground. When perched, it usually blends into its surroundings extremely well.

Gossamer-wing Butterflies

FAMILY LYCAENIDAE

These are small to medium-sized butterflies that are often brightly colored with iridescent blues, reds, and oranges. Their antennae are dark with white rings. Adults usually rest with their wings spread and fly close to the ground in a rapid, but somewhat jerky, fashion.

Hairstreaks—Subfamily Theclinae

Hairstreaks are small to medium-sized butterflies that range in color from iridescent blues and greens to dull browns and grays. Most species possess thin, rather long tails and have a habit of rubbing their hindwings back and forth while feeding, a behavior believed to draw a predator's attention to the butterfly's less vulnerable rear end rather than to its head.

Great Purple Hairstreak *Atlides halesus*

Wingspan, 1.25–1.6 inches. One of our most outstanding butterflies, sometimes known as Great Blue Hairstreak, it is the largest of our hairstreaks. It is triangular in shape when perched and possesses an overall purplish gray color with metallic red spots at the base of the wings. Its two long, twisted, and threadlike tails and bright red abdomen are unique. Dorsally, males are a brilliant iridescent blue color, usually evident only in flight; females are somewhat duller.

SIMILAR SPECIES: No other West Texas butterfly even casually resembles this lovely creature.

WHEN AND WHERE: Although it is widespread in occurrence, usually from March to December, it cannot be expected with regularity. It may be found nectaring on a variety of flowering plants (especially white flowers), but its larval foodplants are limited to various species of mistletoe that occur from the desert lowlands into the mountains.

REMARKS: One would think that this gorgeous creature is tropical in origin, but it occurs throughout the southern United States, from coast to coast, and it commonly wanders north to New York, the Midwest, and Oregon.

Juniper Hairstreak *Callophrys gryneus*

Wingspan, 0.9–1.2 inches. This is an extremely variable butterfly. The underparts of a typical individual reveals a bright green hindwing, crossed by a broad rust-colored line edged with white, indistinct black submarginal spots, and a long tail. The forewing is reddish brown, crossed by a white line aligned with that of the hindwing, and has a greenish base.

SIMILAR SPECIES: Two other green hairstreaks are similar. The Olive Hairstreak, now considered a subspecies of Juniper Hairstreak, differs on the hindwing by having two jumbled, rather than one straight, white-and-rust lines that are not aligned. The Sandia Hairstreak is tailless and lacks the black submarginal spots.

WHEN AND WHERE: Juniper Hairstreaks can be common in spring, especially in the Davis Mountains, nectaring on a wide variety of flowering plants. They can sometimes be flushed by tapping juniper foliage, their larval foodplant.

REMARKS: The West Texas Juniper Hairstreak is now listed as subspecies *siva,* while the Olive Hairstreak is subspecies *gryneus.*

Although fresh Juniper Hairstreaks are easily identified, worn individuals can be difficult. However, in our area, any greenish hairstreak with tails will be this species.

Gray Hairstreak *Strymon melinus*

Wingspan, 1–1.25 inches. Sometimes known as Common Hairstreak, because of its abundance throughout most of North America, it comes in a variety of shades, from slate gray to brownish. The pattern on the underwing is this butterfly's most distinct feature: two sets of narrow white, black, and red jagged lines arise from the upper orange-red spot (one of two) near the tail; one line runs along the margin, and the other crosses the hindwing. A dorsal view shows the larger red spots, with black bases, against a blackish or slate gray background. The abdomen of males is orange.

SIMILAR SPECIES: At least two other hairstreaks possess similar features: Poling's Hairstreak and Red-lined Scrub-Hairstreak. The rare Poling's Hairstreak is all-brown below with a white-edged black postmedian line that forms a small **W** on the hindwing near the tail. The rare Red-lined Scrub-Hairstreak is light gray-brown below and has a jagged red line with a white outer edge on the hindwing, arising from two orange-capped black spots near the tail.

WHEN AND WHERE: Although the Gray Hairstreak is never found in great numbers, it occurs from the Rio Grande floodplain to the mountain highlands. It usually is found at flowering plants, including all the dominant nectar producers, such as acacias and mimosas, beebrush, kidneywood, poreleaf, and even creosotebush. Larval foodplants for this hairstreak are extremely varied, although those in the pea and mallow families may be utilized most often.

REMARKS: Gray Hairstreaks are well-known hill-toppers; males perch up high during morning and late afternoon hours, waiting and watching for passing females.

Blues—Subfamily Polyommatinae

Blues are tiny to small butterflies that vary in color by sex; the dorsal side of males is bright, showy blue, while that of females is dull brown to gray. Their flight is often low, erratic, and up and down, and may be difficult to follow.

Western Pygmy-Blue *Brephidium exile*

Wingspan, 0.5–0.75 inch. Generally regarded as the world's smallest butterfly, it is a lovely creature that is readily identified from below by the golden brown color of the forewings and the grayish hindwings with a row of black marginal spots but no red submarginal band.

SIMILAR SPECIES: No other tiny butterfly in West Texas can be confused with this little gem.

WHEN AND WHERE: Although it may occur almost anywhere, the Western Pygmy-Blue is more numerous in the lowlands and on lower, open slopes of the mountains, often occurring in disturbed areas. In winter, it can be the single most common species, nectaring on whatever flowers it can find. Larval foodplants include a variety of goosefoot species, including pickleweed, saltbushes, and seepweeds.

REMARKS: This butterfly is often overlooked because of its tiny size, but it will usually allow close-up observations when approached slowly, if you do not allow your shadow to frighten it off.

Marine Blue *Leptotes marina*

Wingspan, 0.8–1.15 inches. This butterfly, also known as Striped Blue, is rather easily identified by the multitude of grayish, oblong circles against a tan (females) to purplish (males) background, and a pair of black dots, edged with orange, near the tail. A dorsal view shows a uniform lavender-blue pattern, brighter at the base.

SIMILAR SPECIES: It can be confused with only two other blues, Cassius and Ceraunus. The very rare Cassius Blue is smaller and whitish below and has broken black lines. Ceraunus Blue shows a ventral row of dark dashes on both wings and two small, round black spots on the margin of each hindwing.

WHEN AND WHERE: Marine Blues are widespread, especially near moist areas such as springs and streams, and normally are found only from March through September. Larval foodplants include a wide assortment of legumes, such as acacias, daleas, lotus, indigobush, and mesquites.

REMARKS: This little butterfly is often found in large numbers at choice nectar plants. It is also known for its extensive northward movements in fall, sometimes traveling, probably assisted by wind currents, more than five hundred miles.

Reakirt's Blue *Hemiargus isola*

Wingspan, 0.75–1 inch. This tiny butterfly is readily identified from below by the five large black submarginal spots over a black bar on the forewing, as well as the numerous black spots and whitish rim of the hindwing. A dorsal view shows an overall brownish (females) to blue (males) color with a large and a small black spot at the trailing edge of the hindwing.

SIMILAR SPECIES: It can only be confused with the Ceraunus Blue, which lacks the five prominent black spots on the ventral side of the forewing.

WHEN AND WHERE: Reakirt's occurs year-round and generally at all elevations. It can be common in the lowlands in winter, nectaring on whatever flowering plants it can find, and it can also be common in the mountains from April into October. Larval foodplants include a variety of legumes, such as acacias and mimosas, clovers, daleas, indigobush, and mesquites.

REMARKS: Reakirt's is pronounced "Re-a-kurts."

Reakirt's Blue tend to be loners, so they are sometimes called "Solitary Blue." Because of its southern affinity, some authors call it "Mexican Blue."

Spring Azure *Celastrina ladon*

Wingspan, .75-1.3 inches. This is one of the most variable of butterflies. It is somewhat larger than our other blues, showing bright violet-blue upperparts in flight; when resting, it reveals white, gray, or silver underparts flecked with rows of black spots, with white margins on males and black margins on females.

SIMILAR SPECIES: Only the rare (in our area) Western Tailed-Blue could possibly be confused with the tailless Spring Azure; a close look at Western Tailed-Blue will reveal a tiny tail and two orange tail spots.

WHEN AND WHERE: Spring Azure is a highland species, flying primarily in spring, although it can also occur after rainy periods in late summer and fall. The larva can use a wide variety of foodplants, including blossom buds of various flowering trees and shrubs, such as those of the Texas madrone.

REMARKS: It is one of the earliest butterflies to emerge in spring and has an extremely broad range, from Alaska southward to the mountains of Central America.

Metalmarks

FAMILY RIODINIDAE

Members of this family are small to medium sized, and most possess bands of shiny, metallic-like scales. The front legs of males are reduced in size and not used for walking. Females possess three pairs of walking legs. Adults rest and feed with their wings spread.

Fatal Metalmark *Calephelis nemesis*

Wingspan, 0.75–1 inch. This is a brownish to dark gray-brown butterfly with a band of darker brown across the center of each wing. It also possesses a thin metallic marginal line and irregular checkered fringes, and males have pointed forewings. A ventral view shows a solid blue metallic marginal line against an orange-brown background.

SIMILAR SPECIES: It is rather easy to confuse this species with the somewhat larger Rawson's Metalmark, which occurs in similar habitats. To distinguish these look-alikes for sure, examine the underwing pattern. The metallic margin line is broken on Rawson's but solid on Fatal. The very rare (in our area) Rounded Metalmark shows rounded forewings and an indistinct checkered fringe.

WHEN AND WHERE: Fatal Metalmarks occur at all elevations, but are most numerous in the lowlands and in the lower mountains. They have been recorded year-round but can only be expected from March through November, usually on or near clematis, its larval foodplant. The similar Rawson's Metalmark utilizes various *Eupatorium* species.

REMARKS: The appearance of this species can vary from a chocolate brown to a deep orange color.

Mormon Metalmark *Apodemia mormo*

Wingspan, 0.85–1.25 inches. This lovely butterfly perches with widespread wings, showing numerous bright white and black spots and slashes against an orange background. The base of the wing is reddish, and the fringes are well checkered. The underwing pattern shows larger white patches against a deep orange-brown background. Females are more reddish colored than males.

SIMILAR SPECIES: Although several other butterflies possess similar colors, such as Palmer's and Hepburn's Metalmarks and Texan Crescent, none shows large, bright white spots against an orange background.

WHEN AND WHERE: Mormon Metalmarks are never common, but a few can usually be found with some searching in desert habitats at middle elevations from April through October. Larval foodplants include various buckwheat and ratany species, especially range ratany.

REMARKS: This is a butterfly of Western affinity, where it received its common name. It is a swift flier and often perches with its head up or down in bright sunlight. It also occurs on hilltops, where it usually perches with half-open wings that are slowly fanned in jerky motions.

Palmer's Metalmark *Apodemia palmeri*

Wingspan, 0.75–1 inch. This is a gorgeous species with white blotches against a black, brown, and orange (especially at the wing base) background on the upper side, large white blotches against a pale peach background below, and black-and-white checkered margins.

SIMILAR SPECIES: Chisos and Mormon Metalmarks possess similar features, but the Mormon Metalmark is larger. Dorsally, the white blotches of Palmer's are brighter and stand out more against a black background. The Chisos Metalmark is black and orange with no dorsal white blotches, and a ventral view shows black blotches against a white background on the hindwing.

WHEN AND WHERE: The Palmer's Metalmark can be fairly common at low and middle elevations, especially after rainy periods. Mesquite is its principal larval foodplant.

REMARKS: Although adults often perch on their larval foodplant, they take nectar from a wide variety of flowering shrubs and herbs.

Brushfooted Butterflies

FAMILY NYMPHALIDAE

All members of this family possess reduced forelegs, although they have few other characteristics in common. The short forelegs are covered with spines, especially in the males. They usually rest with wings folded but often bask spread-winged.

Snouts—Subfamily Libytheinae

Snout butterflies possess a long snout that usually is readily apparent. When perched, they mimic dead leaves. Unlike other brushfoots, female snouts possess six visible walking legs while the male's front legs are reduced and not used for walking. The flight pattern is strong but fluttery.

American Snout *Libytheana carinenta*

Wingspan, 1.4–1.8 inches. One of the most abundant and distinct butterflies of West Texas, it is readily identified by its rather obvious "snout," forward-thrust labial palps. Its underwings are a mottled, brownish gray, while the forewings possess an orange base, a white patch surrounded by black, and a square tip. When sunning with wings spread, it reveals orange-brown stripes against a blackish background and a series of white spots across each forewing.

SIMILAR SPECIES: No other butterfly possesses similar features.

WHEN AND WHERE: Snouts can be found year-round, although they are far more numerous in summer than in winter. Even though they are widespread and can be found in almost any habitat, they usually are far more numerous in the vicinity of hackberries, their principal larval foodplants. They also spend much time on honey mesquites.

REMARKS: Snouts are famous for their occasional migrations when millions sometimes emerge during the fall, usually following a summer drought, and fly in enormous flocks.

Heliconians and Fritillaries—Subfamily Heliconiinae

These are medium to large butterflies, several of which are extremely colorful. Adults possess elongated wings and long antennae. Their flight is often a slow flutter, but they can fly swiftly when threatened. Many are distasteful or mimic species that predators shun.

Variegated Fritillary *Euptoieta claudia*

Wingspan, 1.75–2.5 inches. This is an orange-brown to tawny brown butterfly, readily identified by the numerous black zigzag lines on the forewings and a row of black submarginal dots and black marginal lines on both wings. There are no silver spots on the ventral side.

SIMILAR SPECIES: There are two similar species. The Gulf Fritillary possesses longer wings, with bright orange above and bright silver spots on the underside. The rare Mexican Silverspot's dorsal side is brownish orange above, and the underside possesses cloudy black marks on the basal half and elongated silver spots on the hindwings.

WHEN AND WHERE: One the area's most common large butterflies for most of the year, Variegated Fritillaries prefer open sites with scattered flowers. Larval foodplants include several violet species and the spread-lobe passion flower.

REMARKS: This butterfly usually flies low over the ground with shallow wing beats and while nectaring, often flutters over a flower rather than alighting.

True Brushfoots—Subfamily Nymphalinae

These are mostly medium-sized butterflies that possess diverse shapes and patterns and vivid colors. Most have crescent-shaped spots on their wing margins.

Bordered Patch *Chlosyne lacinia*

Wingspan, 1.4–2 inches. Although this is one of the most variable species, the basic pattern of a broad transversal orange-to-yellow line across the black hindwings is typical. The underside is similar except that the yellow crescent markings on the wing margins are more obvious.

SIMILAR SPECIES: There are no similar species that one is likely to encounter in West Texas.

WHEN AND WHERE: The Bordered Patch can occur anywhere from the floodplain of the Rio Grande to the mountain highlands, although it is most numerous in low and middle elevations. It can be abundant in midsummer when nectar plants are readily available. Larval foodplants include numerous species of composites, although goldeneye and sunflowers are most popular.

REMARKS: Adults take nectar from flowers, especially white and yellow ones, and males take nutrients from dung, carrion, and mud.

Elada Checkerspot *Texola elada*

Wingspan, 0.85–1.25 inches. This little butterfly is best identified dorsally by its orange-and-black checkered pattern. A ventral view reveals a checkered black-and-white pattern with a red-orange marginal band. Its rather long wings are also noticeable.

SIMILAR SPECIES: The Tiny Checkerspot, that is similar in size, is distinguished ventrally by a white instead of a red-orange marginal band.

WHEN AND WHERE: Elada Checkerspots can be found at almost all elevations from March through October. They can be numerous at middle elevations in midsummer if rains have produced larval foodplants: members of the acanthus family, especially tubetongues.

REMARKS: Elada and Tiny Checkerspots often fly together. They can be abundant in arroyos of the mountain lowlands, especially when kidneywood shrubs are flowering.

Elada Checkerspots often obtain nutrients from animal carcasses. On September 3, 1988, I discovered several pieces of a deer carcass, an obvious lion kill, which were literally covered with butterflies. Of nine species identified, 120+ were Elada Checkerspots, 30+ Tiny Checkerspots, 25+ Reakirt's Blues, 5 Queens, 4 Golden-headed Scallopwings, 4 American Snouts, 3 Marine Blues, one Western Pygmy-Blue and one streaky-skipper.

Texan Crescent *Phyciodes texana*

Wingspan, 1.25–1.75 inches. This butterfly perches with open wings and is readily identified by its overall blackish color with a rather obvious line of white spots across the hindwings. There are also reddish blotches at the base of the wings. The underside, though rarely seen, is orange and brown and crossed with a line of white crescent spots.

SIMILAR SPECIES: No other butterfly in our area possesses a similar dorsal pattern.

WHEN AND WHERE: The Texan Crescent is most numerous along mid-elevation arroyos but occurs almost anywhere from March through November. Although it can also be found in open areas, it may seek shady sites in midday. Larval foodplants include several members of the acanthus family, such as anisacantha, ruellia, and tubetongue.

REMARKS: Texan (not Texas) Crescents rarely fly far; instead, they usually patrol small areas where the same individual can be found on several consecutive days. Males often perch on the ground or on shrubs, waiting for a passing female.

Texan Crescents are extremely territorial. They chase almost anything moving through their territory, whether it is another butterfly, any other insect such as a fly or bee, or even a bird. But after a short chase, they will return to the same small territory.

Mourning Cloak *Nymphalis antiopa*

Wingspan, 2.7–3.4 inches. A large and gorgeous butterfly, it is easily identified by its purplish black wings with broad yellowish margins.

SIMILAR SPECIES: No other West Texas butterfly can be confused with this lovely creature.

WHEN AND WHERE: Although the Mourning Cloak can be found almost anywhere, it prefers moist areas such as streamsides and ponds. Larval foodplants include willows, cottonwoods, and hackberries.

REMARKS: The name was derived from its gold-edged dark color, like a cloak worn by mourners. This butterfly overwinters as an adult and may fly at any time of the year.

Mourning Cloaks often overwinter in the adult stage, even in more northern states where snow can cover the ground. But on sunny days when temperatures may reach in the mid-40s or 50s, it would not be unexpected to see one of these hardy butterflies flying about. They have a habit of clicking their wings in flight or when alighting.

American Lady *Vanessa virginiensis*

Wingspan, 1.75–2.25 inches. This lovely creature usually perches with wings open, revealing a colorful orange and black pattern. The black wing tips also contain a number of white spots, and a tiny white dot is usually present in the orange cell below the black wing tips. A ventral view reveals two large submarginal eyespots, linked together by a black bar, on each hindwing.

SIMILAR SPECIES: Both Painted and West Coast Ladies can also be found in West Texas. Painted Ladies show less contrast below, and the ventral hindwing eyespots are much smaller and less noticeable than those of American Ladies. West Coast Ladies are slightly smaller with squared forewings, and the three isolated eyespots visible on the upper hindwings are black with a blue center.

WHEN AND WHERE: The American Lady can occur year-round, although it is most numerous from March through October. It seems to have a preference for open areas with low vegetation. Larvae utilize numerous members of Asteraceae, such as yarrow, huisache-daisy, ragweed, and aster, as foodplants.

REMARKS: Many authors refer to this butterfly as American Painted Lady. Adults sip flower nectar and occasionally aphid honeydew.

Red Admiral *Vanessa atalanta*

Wingspan, 1.75–2.5 inches. This is a striking creature, showing velvet-black wings with a bright red marginal band; another red band crosses the forewings. The black wing tips are marked by several large square white dots. The underside reveals mottled browns and blacks on the hindwing; the forewings look similar to their dorsal view.

SIMILAR SPECIES: No other species possesses such distinct features. The smaller Bordered Patch is also black with a red, orange, or yellowish band across the wings but lacks the marginal band.

WHEN AND WHERE: Although Red Admirals can be found year-round, they are more numerous at lower elevations in early spring and fall, and more numerous at higher elevations in late spring and summer. Their larval foodplants include various nettle and false nettle species.

REMARKS: Adults hibernate. They feed on flower nectar, sap, fruit, and dung. Their flight pattern is extremely fast and difficult to follow, but they commonly return to an established territory.

On the early morning of April 25, 2001, I found a total of 42 Red Admirals sunning on grape and cottonwood leaves at Rio Grande Village in Big Bend National Park. With wings spread to absorb warmth from the sunlight, they provided a marvelous black and red pattern among the bright green foliage.

Common Buckeye *Junonia coenia*

Wingspan, 1.65–2.4 inches. It usually perches with spread wings, making identification easy. A dorsal view shows large purple, tan, and black eyespots against a brownish background. There are also two red bars, edged with black, near the base of each forewing. The undersides are not so distinctly marked: the hindwings are a brownish color, while the forewings possess one large eyespot, a whitish or yellowish blotch, and a red patch broken by a white line near the base.

SIMILAR SPECIES: Only the rare Dark Buckeye subspecies of the Tropical Buckeye can be confused with this species. However, the Dark Buckeye is much darker, even blackish, with only a hint of the Common Buckeye's eyespots above, and with a somewhat lighter band below.

WHEN AND WHERE: Although Common Buckeyes are widespread in occurrence, they are rarely found in abundance; loners prevail. They can be expected year-round in the lowlands and from March to November at higher elevations. Larval foodplants consist of numerous members of the Acanthaceae, Plantaginaceae, Scrophulariaceae, and Verbenaceae families.

REMARKS: The relationship between the various species of Buckeyes is not well understood; the Dark Buckeye may be a distinct species. Males are not generally territorial but perch on the ground or low plants waiting for passing females.

Admirals and Relatives—Subfamily Limenitidinae

These butterflies are mostly large species characterized by a flap-and-glide flight. Adults do not normally visit flowers and usually perch head-down on tree trunks with wings spread. They may produce "click" sounds when flying.

Red-spotted Admiral *Limenitis arthemis*

Wingspan, 2.25–3.4 inches. This is a large, almost swallowtail-sized butterfly without the swallowtail. It is most often found in flight, where it reveals an overall blackish color except for the trailing wing edges, which are deep blue with black bands above. A ventral series of deep red submarginal spots and black-and-blue shallow crescent spots on the margins can be seen when the butterfly is perched, as it often is. There are also a number of red blotches on the base of each wing.

SIMILAR SPECIES: The Pipevine Swallowtail possesses a similar dorsal pattern but has tails.

WHEN AND WHERE: Red-spotted Admirals are a mountain species and rarely occur below the moist, wooded canyons. They rarely fly before late April or after October. Larval foodplants consist of members of the rose and willow families, such as chokecherry and hornbeam, and cottonwoods and willows, respectively.

REMARKS: This species, previously known as Red-spotted Purple, mimics the toxic Pipevine Swallowtail, thus gaining protection from various predatory birds. Males perch on shrubs and trees, watching for passing females. Adults feed on flower nectar, sap, fruit, carrion, dung, and honeydew.

California Sister *Adelpha bredowii*

Wingspan, 2.25–3 inches. This striking butterfly is readily identified from above by the large orange-red, somewhat squared spot at the tip of each forewing and a series of large white blotches that run across the black-brown wings. When seen from below, the hindwing possesses a matrix of colors: tans, oranges, blues, and whites. The forewing undersides show the large orange-red dots and a series of white dots.

SIMILAR SPECIES: No other large West Texas butterfly possesses the unique colors and pattern of the California Sister.

WHEN AND WHERE: This is another mountain species that can be found flying throughout the woodlands and especially patrolling up and down the canyons; it rarely visits the lower canyons. It can usually be found from March into November. Larval foodplants include various oaks, such as Emory oak.

REMARKS: It can often be approached quite closely when it is perched at wet and muddy places.

Common Mestra *Mestra amymone*

Wingspan, 1.4–1.75 inches. In flight, this little butterfly shows an all-white or gray upper side with orange on the trailing portion of the hindwings. A ventral view reveals lines of white blotches against a pale orange background.

SIMILAR SPECIES: No other West Texas butterfly can be confused with this lovely creature.

WHEN AND WHERE: Although Mestras can occur at all elevations, they are most common in canyons at middle elevations. They can be found as early as May but become more numerous by middle to late summer, especially after rainy periods. Larval foodplants are limited to euphorbs, such as noseburn species.

REMARKS: One of the daintiest of butterflies, it has a slow, sailing flight that can often be detected from a considerable distance. Many authors refer to this species as Amymone.

The Common Mestra appears in late spring, slowly flying in the cool, shaded canyons. It never flies far when disturbed. In the lower Big Bend Country it has several broods each year. However, this species is known to swarm northward in summer.

Leafwings—Subfamily Charaxinae

Leafwings are medium to large butterflies that possess bright, colorful upper wings but dull underwings, so they are well camouflaged when perched. They are also extremely fast flying. They generally possess a stout body and large palps.

Tropical Leafwing *Anaea aidea*

Wingspan, 2.25–2.45 inches. This large butterfly possesses a rich orange-red color with narrow blackish margins on its dorsal side and mottled gray-brown undersides. When it is perched, its scalloped hindwing outline and tail-like projection can be seen.

SIMILAR SPECIES: The less numerous and slightly larger Goatweed Leafwing is best distinguished by the smooth outline of its hindwing margins.

WHEN AND WHERE: Although widespread, Tropical Leafwings are most numerous from March through October in the mountain canyons and less abundant in lower wooded arroyos. Larval foodplants are limited to various crotons.

REMARKS: This is a fast-flying butterfly that investigates each passing butterfly or almost any other passing creature. Usually perched on a tree or tall shrub, it will zoom out to investigate and then swiftly fly about before landing. Once perched, unless seen landing, it is difficult to locate. Adults feed on sap.

Emperors—Subfamily Apaturinae

These are mid-sized colorful butterflies with a stout body. Adults are fast fliers and often perch on high foliage, but will also perch on the ground. They are quite territorial, usually returning to a favorite perch when disturbed. They take nectar from flowers and obtain nutrients from sap, fruit, and dung.

Empress Leilia *Asterocampa leilia*

Wingspan, 1.5–2 inches. It often perches with wings spread, showing a rich red-brown color on a fresh specimen. The wing tips are black with several white spots, and both forewings and hindwings possess a series of eyespots. A ventral view reveals two solid brown discal bars and two moderately large submarginal eyespots on the forewing and a mottled grayish brown-and-blue hindwing with several small eyespots.

SIMILAR SPECIES: Both the Tawny and Hackberry Emperors are similar. Tawnys lack any forewing eyespots and possess two prominent black discal bars. Hackberry Emperors possess forewing eyespots but only one solid black discal bar.

WHEN AND WHERE: Empress Leilias can be commonplace along low and mid-elevation arroyos and appear even in the wooded mountain canyons. They are usually present from March to mid-November; males can often be found occupying territories two hundred to four hundred feet apart. All of the hackberry butterflies utilize various species of hackberry for their larval foodplants; Empress Leilias utilize spiny hackberry.

REMARKS: Several subspecies of Emperors can occur that appear slightly different from the three key species. All the males spend the majority of their time perched or patrolling in confined territories in arroyos, chasing passing butterflies. They may even investigate you in passing. Adults feed on sap and dung and only occasionally take flower nectar.

Satyrs—Subfamily Satyrinae

Satyrs, small to medium-sized butterflies, are predominantly brown with one or several marginal eyespots. Their costal vein has a swollen base. Their flight is low to the ground and exhibits a distinct bouncing or "hopping" pattern.

Red Satyr *Megisto rubricata*

Wingspan, 1.5–1.75 inches. It can usually be identified, even from a distance, by its bouncy and erratic flight. When perched with closed wings, it displays gray-brown undersides crossed with dark lines, a reddish patch on the forewing, and two large eyespots, one on the forewing and one on the hindwing. A dorsal view reveals a large eyespot (black with a yellow edge) near the tip of each forewing and a deep reddish patch on the otherwise dark brown wings.

SIMILAR SPECIES: Two other satyrs in this region are similar. Mead's Wood-Nymph resides in the mountain woodlands, where it stays close to pine trees. It shows a reddish flush on dark brown wings like Red Satyr's, but it has two prominent eyespots rather than one on each forewing. Canyonland Satyr (fairly common during springtime in the Davis Mountains) lacks eyespots but possesses a large grayish patch with three shiny black spots on the outer edges of its hindwings.

WHEN AND WHERE: Another of the mountain species that flies from March through October, it rarely is found below the woodlands. It frequents shady areas, flying low over the ground. Larval foodplants include several grasses.

REMARKS: Adults rarely take nectar. Males patrol shaded areas all day in their search for a mate.

Monarchs—Subfamily Danainae

These are large, orange-and-black butterflies that utilize milkweeds (Asclepiadaceae) as foodplants and nectar sources. The toxic ingredients obtained make them distasteful to predators, typically providing them vital protection. Their flight pattern is slow but strong.

Monarch *Danaus plexippus*

Wingspan, 3.5–4 inches. Probably the best known of all butterflies, this large creature is overall orange with noticeable black veins. The broad black margins of both wings contain many white spots. A ventral view reveals yellow-orange "windows" between the black veins.

SIMILAR SPECIES: The smaller Queen lacks both the heavy black veins above and the orange "windows," but possesses numerous white dots on the wing tips. The Viceroy is also similar but possesses a black line across the hindwings and a single row of white spots in the black marginal band of the forewings.

WHEN AND WHERE: Monarchs are most numerous during their fall migration, but they can be found almost any time of year and almost anywhere. Larval foodplants are limited to a variety of milkweeds, including several common species such as Texas milkweed, antelopehorn, and both arroyo and desert twinevines.

REMARKS: Monarchs migrate in spring and fall, although no single individual makes the entire round-trip journey. Adults may take nectar from a wide assortment of flowers.

Queen *Danaus gilippus*

Wingspan, 2.6–3.4 inches. Only slightly smaller than Monarchs, Queens are orange-brown with bold black veins and numerous scattered white spots near the wing tips.

SIMILAR SPECIES: Monarchs and Viceroys are similar; see the previous descriptions.

WHEN AND WHERE: Queens are most numerous at middle and lower elevations in summer and fall, but they can occur almost anywhere and at any time. Some years they are abundant over the desert lowlands. Queen foodplants, as for Monarchs, are limited to milkweeds.

REMARKS: Adults regularly roost communally. This butterfly is one of the earliest to fly in the mornings.

Queens can be superabundant during late summer and fall when summer rains have produced an adequate nectar source. I have found their numbers exceeding six dozen per mile along the highways in the Big Bend Park area.

Skippers

FAMILY HESPERIIDAE

Skippers are an extremely diverse group of tiny to medium-sized butterflies that are mostly brown, gray, yellow, or orange in color. Both sexes possess six well-developed walking legs. The tip of each antenna is bent downward to include a thicker, flattened extension (apiculus) but is not knobbed as it typically is in butterflies. Most individuals are very strong fliers, and their flight pattern is usually jerky or "skipping," providing them with their common name.

Spread-wing Skippers—Subfamily Pyrginae

Adults typically rest with their wings spread, providing the observer an excellent view of their dorsal side. Males never possess a stigma but often have a costal fold.

Golden Banded-Skipper *Autochton cellus*

Wingspan, 1.6–1.85 inches. This skipper is readily identified by a solid, broad yellow-gold band across each black-brown forewing. It also has a small white bar near the wing tip and a white ring on each antenna below the club.

SIMILAR SPECIES: The rare Chisos Banded-Skipper (see page 61) is found only in the Chisos Mountains in the United States. It possesses a narrow white band across each forewing, as well as broad white margins on each hindwing.

WHEN AND WHERE: This distinctly marked skipper can usually be found in the mountains, especially in moist areas, from March through November. Its larval foodplants are limited to a number of peas, most notably Wright bean.

REMARKS: It is extremely territorial and often perches on shrubs and rocks at eye level. It is not easily startled, often allowing close inspection.

Northern Cloudywing *Thorybes pylades*

Wingspan, 1.25–1.65 inches. This is a tan to brown skipper with separate, white glassy spots on the outer portion of each forewing. These spots are small, nonaligned, and triangular. The wing fringes are checkered.

SIMILAR SPECIES: The Desert Cloudywing is dark brown with two dark bands on the hindwings, a mottled grayish marginal band, and uncheckered hindwing fringes. The hindwings of Desert Cloudywings are longer, a useful first clue in distinguishing these similar-sized cloudywings.

WHEN AND WHERE: Northern Cloudywings are most numerous at higher elevations, although they can also occur in the lower foothills. Larval foodplants include several members of the legume family; Texas senna is especially favored.

REMARKS: Unlike many skippers, Northern Cloudywings often perch on the ground, sometimes with their wings held half open.

Northern Cloudywings can be extremely variable, but males are easily identified: they are the only cloudywings with a costal fold on the forewing.

Golden-headed Scallopwing *Staphylus ceos*

Wingspan, 1–1.1 inches. This spread-winged skipper has blackish brown wings and a noticeably golden-colored head with rather long palps. The forewings are rounded, and the hindwings are slightly scalloped.

SIMILAR SPECIES: Three similar sootywings—Common, Mexican, and Saltbush—are all about the same size and possess dark wings but lack the golden head. In the Davis Mountains, watch also for the Orange-headed Roadside-Skipper, which perches with folded wings.

WHEN AND WHERE: It is most common in spring and summer but is also seen irregularly in fall, mostly at middle and higher elevations. This species tends to perch on or under leaves. Larval foodplants include members of the goosefoot family.

REMARKS: Males patrol arroyos, watching for mates, and sometimes fly in shady areas near streams.

Although fresh Golden-headed Scallopwings are easily identified, worn individuals and females can be difficult. They may lack a golden head, so identification of this spread-winged skipper may require a careful examination. Some gold is usually present.

Texas Powdered-Skipper *Systasea pulverulenta*

Wingspan, 1–1.2 inches. A dorsal view of this butterfly reveals a mottled pattern of brown, black, and rust colors, with a narrow band of white that is aligned where it crosses both wings. The wing base is blackish, while the lobed outer edge of the hindwing is rust colored.

SIMILAR SPECIES: The rare (in our area) and somewhat larger Arizona Powdered-Skipper lacks the contrasting pattern and has longer wing lobes, and the white line is not aligned across the wings.

WHEN AND WHERE: This skipper flies most of the year and can be expected from the lowlands into the lower mountains. Larval foodplants consist of several members of the mallow family, including various *Abutilon* species.

REMARKS: Texas Powdered-Skippers seem to have an affinity for moist areas. Although they can never be expected with certainty, I have found them most often at springs and in cooler mountain canyons.

Rocky Mountain Duskywing *Erynnis telemachus*

Wingspan, 1.2–1.8 inches. A dorsal view of this duskywing reveals an overall pale brown-gray individual with dark and bluish patches and a series of four small white glassy spots on the forewings. The hindwing fringes are dark. A ventral view shows two small pale spots along the leading edge of the hindwing.

SIMILAR SPECIES: Only the Meridian Duskywing can be confused with the Rocky Mountain Duskywing but is overall blackish in color with little contrast, and the hindwing fringes are pale-tipped on fresh individuals.

WHEN AND WHERE: This is a mountain species most often found in cooler canyons, where it often perches on the ground or in wet areas. It flies from March through September. Larval foodplants include a number of oak species.

As might be expected from its name, this spread-winged skipper reaches the southern limits of its range in the West Texas mountains. It is a more common resident in the oak woodlands in the mountains of New Mexico, Arizona, Utah, and Colorado.

REMARKS: It looks very much like the slightly smaller Juvenal's Duskywing but without white margins on the hindwings.

Funereal Duskywing *Erynnis funeralis*

Wingspan, 1.35–1.75 inches. This spread-winged skipper is easily recognized by its relatively square, blackish hindwings with broad white fringes. The forewings, longer and more pointed than those of other duskywings, are mottled dark brown and black, with a brown patch and a row of small white glassy hyaline spots.

SIMILAR SPECIES: At least three West Texas skippers can be confused with the very dark Funereal Duskywing. Mournful Duskywing also possesses white fringes but has broad forewings, brown instead of black hindwings, and white spots inside the white fringes. Juvenal's Duskywing has shorter forewings with buff spots, often hidden by long gray hairs, and the underside has two round light spots on the leading edges of the hindwings. Drusius Cloudywing possesses larger white spots on the forewings, and the white hindwing fringes are scalloped.

WHEN AND WHERE: Funereal Duskywings are widespread and can be found most easily from March through October. Preferred habitats range from arroyo bottoms to open woodlands and desert flats. These butterflies often perch on leaves at eye level. Larval foodplants include a wide variety of legumes, such as vetch and lotus species.

REMARKS: As for the Mourning Cloak, the Funereal Duskywing's name was derived from its dark coloration with light-colored fringe.

Common Checkered-Skipper *Pyrgus communis*

Wingspan, 1–1.25 inches. This little checkered-skipper shows an extremely variable pattern above and below. A dorsal view reveals blackish wings with numerous white patches, including a series of square patches across both wings, numerous additional square spots, especially near the wing tips, and a checkered fringe on the forewing. The body of some individuals, especially males, is covered with greenish hairs. The underside is similar but paler and crossed by four rows of olive spots outlined with dark scales.

SIMILAR SPECIES: Small and Desert Checkered-Skippers possess like features. The rare Small Checkered-Skipper is tiny and brown, without dark spot outlines, and has broad white fringes. The Desert Checkered-Skipper is blackish or dark brown, often has bluish hairs but no black spot outlines, and has a small round white spot above the square basal spot. Its underwings are a gray or silvery color.

WHEN AND WHERE: Common Checkered-Skippers can be found almost anywhere year-round. It prefers open areas where it usually perches on the ground. Larval foodplants consist of various members of the mallow family.

REMARKS: This species is nonterritorial, roaming about in search for a mate. Adults feed on flower nectar and sap.

Common Streaky-Skipper *Celotes nessus*

Wingspan, 0.8–1 inch. This tiny skipper is marked with orange-brown and dark brown streaks, giving it a pleated appearance. The hindwings are scalloped and possess checkered fringes.

SIMILAR SPECIES: The Scarce Streaky-Skipper, which can be fairly common in the south but less numerous northward, is so similar that the only sure way to distinguish these two species is by genitalic differences. They rarely are distinguished in the field by the shape of the tiny white dots across the forewings dorsally: Scarce Streaky-Skipper shows a line of jumbled white dots while Common shows a straight line of white dots.

WHEN AND WHERE: Both streaky-skippers are widespread and can be found from March through October. However, they are more numerous in spring to midsummer than later in the year. Both nectar on various flowers, and they often perch on the ground. Larval foodplants consist of various members of the mallow family.

REMARKS: Positive identification of the two streaky-skippers requires examination of the genitalia. Scott (1986:498) describes the methods you can use to distinguish the Common Streaky-Skipper from the Scarce Streaky-Skipper.

Common Sootywing *Pholisora catullus*

Wingspan, 0.8–1.25 inches. This spread-winged, glossy black skipper possesses a fairly long rounded wing with numerous white dots. The hindwing fringes are brownish black, not checkered.

SIMILAR SPECIES: Two other sootywings are similar. The rare Mexican Sootywing shows a distinct bluish gray sheen and black-lined veins on the underside of the forewings. The somewhat smaller Saltbush Sootywing, usually found in the lowlands near saltbushes, is brown with a grayish sheen and a row of dark arrowhead marks on the forewings; it also has checkered fringes.

WHEN AND WHERE: Common Sootywings can occur from March through October, primarily at middle elevations. They spend a good deal of time perched on the ground. Larval foodplants include various amaranth, goosefoot, and saltbush species.

REMARKS: Males patrol for females low over the ground in arroyos and in disturbed areas.

The similar Saltbush Sootywing can be abundant in the lowlands. I recorded 13 and 25 individuals flying about two four-wing saltbushes at Cottonwood Campground in Big Bend National Park on March 5, 1999.

Grass- or Closed-wing Skippers—Subfamily Hesperiinae

These are tiny to small creatures that usually perch with their forewings upright over their backs, although they may bask with their wings spread. These skippers also possess a stigma, a patch of raised scent scales, on the upper surface of the male's forewings.

Orange Skipperling *Copaeodes aurantiacus*

Wingspan, 0.75–1 inch. As suggested by its name, this is a tiny skipper with gold-orange wings and very short antennae. A dorsal view reveals a black wing base. The underwings are pale yellow-orange, and the underside of the body is white. Its abdomen extends beyond its wings.

SIMILAR SPECIES: Two other species are similar. The Southern Skipperling, which resides in grassy areas in the lowlands, shows a whitish streak (ray) that runs the length of the underside of its hindwing. The rare (in our area) Tropical Least Skipper has a narrow black border on both wings and a pale ray on the underside of its golden-orange hindwing.

WHEN AND WHERE: Orange Skipperlings are most often found in grassy areas at almost every elevation from March through November. Larval foodplants include a variety of grasses, including Bermuda-grass.

REMARKS: This little skipper is usually found perched at the tip of grasses, with wings partially open, and often appears in pairs. It is easily overlooked due to its small size.

Sachem *Atalopedes campestris*

Wingspan, 1.25–1.55 inches. This is a fairly large tawny orange-and-brown skipper. It is sexually dimorphic: a dorsal view of the male reveals a massive square black stigma against an otherwise tawny orange hindwing; the female shows a pair of yellow bars against a blackish background on the forewing and a square whitish spot near the base of the forewing. The undersides on both sexes are marked with several whitish squares.

SIMILAR SPECIES: The rare Taxiles Skipper is also tawny orange, but the undersides on the female show small pale orange submarginal spots. The male possesses yellow bars but lacks the submarginal spots on the underside of the forewing. The Fiery Skipper is also orange in color but possesses several tiny black spots on the underside of the hindwing.

WHEN AND WHERE: Sachem is widespread and can usually be found from February through November. Larval foodplants consist of numerous grasses, including Bermuda-grass and crabgrass.

REMARKS: Males perch in sunny grassy areas all day, watching for females.

Sachems can appear almost anywhere, from the desert lowlands to the mountain canyons. Because the sexes are dimorphic and they can appear very different in different lights and habitats, they are easily misidentified.

Viereck's Skipper *Atrytonopis vierecki*

Wingspan, 1.25–1.5 inches. The undersides of this skipper are pale brown with lavender-gray overscaling toward the buff-colored, uncheckered margins; two lines of black dots are evident on the hindwing.

SIMILAR SPECIES: Although all the West Texas *Atrytonopsis* skippers—White-barred, Python, Sheep, and Viereck's—have a similar silhouette, Viereck's is the only one with two-toned undersides without checkered margins. The very similar Python Skipper has distinct checkered margins.

WHEN AND WHERE: This species occurs only at low and middle elevations and is most often found nectaring on cactus flowers. It normally flies only in the spring, from March through May. The larval foodplants are unknown, although most entomologists claim that it uses grasses.

The Viereck's Skipper is the most common of the *Atrytonopsis* skippers, but all the species are fairly common during the spring months, especially in Big Bend National Park. However, they are rarely found by late summer and fall.

REMARKS: Males usually perch on twigs or on cactus pads in open arroyos or slopes, awaiting passing females.

Sheep Skipper *Atrytonopis edwardsii*

Wingspan, 1.25–1.5 inches. This skipper perches with its hindwings held high over the rounded forewings. The undersides are brown with a gray overcast, and several round whitish dots are present on the hindwing. The fringes are checkered brown and white. A dorsal view reveals numerous irregular white spots across both wings.

SIMILAR SPECIES: Three other *Atrytonopsis* skippers are similar. White-barred is most alike, with two series of white bars on the hindwing, but the fringes are not checkered. Viereck's Skipper is somewhat smaller and paler; see its description above. The Python Skipper has brown hindwings that are frosted with grayish violet and have numerous inconspicuous white spots; the margins are checkered white-and-brown.

WHEN AND WHERE: The Sheep Skipper can be fairly common in spring and summer but is rare later in the year. This skipper flies at middle elevations, preferring arroyos and sunny slopes. Larval foodplants include a few grasses, such as ryegrass.

REMARKS: Males perch in sunny sites on cliffs and in rocky areas. All the *Atrytonopsis* skippers seek nectar in cactus flowers, often disappearing from sight while nectaring.

Bronze Roadside-Skipper *Amblyscirtes aenus*

Wingspan, 1–1.1 inches. This is a stubby-winged skipper with a brassy glaze and checkered margins. The underside of the hindwing is red-brown at the base and marked with three fairly large square white dots. The underside of the forewing has a transverse row of pale spots.

SIMILAR SPECIES: All the Amblyscirtes skippers are somewhat alike. The Oslar's Roadside-Skipper and Cassus Roadside-Skipper (most common in the Davis Mountains) are most similar. Oslar's is a lighter brown color and lacks the forewing spots; Cassus is orangish with more prominent spots. The Texas Roadside-Skipper is brown with a grayish wash and a vague row of whitish spots across the hindwings, three prominent spots on the forewing, and checkered fringes. The Nysa Roadside-Skipper has mottled gray-and-brown undersides, including a square black central mark on the base of the hindwing.

WHEN AND WHERE: The Bronze Roadside-Skipper flies from March through September and can be found at all elevations. It prefers arroyos, trails, and open rocky sites. Larval foodplants consist of various grasses.

REMARKS: Males perch in rocky arroyos and on open slopes to await females.

West Texas Specialties

The following butterflies normally occur in the United States only within the West Texas region. Many are rare or fly only during brief periods of the year. They are some of the most sought-after species.

Poling's Hairstreak *Satyrium polingi*

Wingspan, 1–1.25 inches. This dark brown hairstreak is marked below with black lines edged with white; the inner line forms a **W** shape. A small orange-and-black spot and a blue patch are present near the tail. This rare hairstreak resides in mountain ranges in West Texas and adjacent New Mexico, Arizona, and northern Mexico. Larval foodplants include gray and Emory oaks.

Sandia Hairstreak *Callophrys mcfarlandi*

Wingspan, 1–1.2 inches. This is a tailless species with golden green undersides and gray-brown to rust-brown upper sides. It occurs in Texas only from the Rio Grande northward in decreasing numbers to the Guadalupe Mountains. Foodplants are limited to various beargrass species that occur on the lower mountain slopes.

Hepburn's Metalmark *Apodemia hepburni*

Wingspan, 0.75–1.2 inches. This butterfly is very similar to Palmer's Metalmark but somewhat smaller. A dorsal view reveals tiny white spots on both wings and a trace of orange at the base. The submarginal spots are black instead of white. It has been recorded in the region only in the Chisos Mountains during July.

Chisos Metalmark *Apodemia nais chisosensis*

Wingspan, 1.1–1.25 inches. This beautiful butterfly is orange, white, and black and flies at middle elevations from April to August. A dorsal view reveals many black lines against an orange background. The underside of the hindwings is whitish with black marks. Green Gulch in Big Bend National Park offers the best site to find this lovely creature.

Chinati Checkerspot *Thessalia chinatiensis*

Wingspan, 1.25–1.5 inches. The undersides of this well-marked Nymphalid are orange with black veins and have a cream-colored submarginal band and reddish spots within a black-rimmed marginal band. Resident in several West Texas mountain ranges, it flies from May through September, usually near its larval foodplant, Big Bend silverleaf.

Chisos Banded-Skipper *Autochton cincta*

Wingspan, 1.2–1.75 inches. This blackish spread-winged skipper is similar to the Golden-banded Skipper but with a very narrow white bar that crosses each forewing. The hindwing margins are also white. It has been recorded in Big Bend Park's Green Gulch on a number of occasions from March to September.

Drusius Cloudywing *Thorybes drusius*

Wingspan, 1.25–1.75 inches. This is a walnut brown spread-winged skipper with a circle of white blotches on the forewings and scalloped white fringes on the hindwings. The underside possesses wavy brown bands on the hindwings. It is reasonably common in the Davis Mountains but rare in the Big Bend region.

Chisos Skipperling *Piruna haferniki*

Wingspan, 0.75–0.9 inch. This little Mexican grass skipper, also known as Hafernik's Skipperling, has been recorded in the United States only in the Chisos Mountains of Big Bend National Park. The undersides are blackish brown with a grayish base. A dorsal view reveals small white spots on a dark brown forewing.

Sunrise Skipper *Adopaeoides prittwitzi*

Wingspan, 0.8–1.1 inches. This appears to be a large version of an Edward's Skipperling, complete with orange-gray undersides, but the Sunrise Skipper possesses black-streaked wing margins on the upper side. The underside has a white ray extending the length of the hindwing. It is fairly common in the Davis Mountain grasslands in spring and summer.

Chisos Giant-Skipper *Agathymus chisosensis*

Wingspan, 1.75–2.25 inches. Regarded as a subspecies of Orange Giant-Skipper (*A. neumoegeni*) by many lepidopterists, it is known only from Big Bend's Chisos Mountains where its larval foodplant—*Agave havardiana*—is common. It is a large orangish skipper with violet-gray scaling; the hindwing is pale gray with a band of smudged black spots along the margins.

Mary's Giant-Skipper *Agathymus mariae*

Wingspan, 1.75–2 inches. This large skipper possesses a dorsal band of orange-yellow spots across both wings, as well as yellow margins. The undersides are pale gray with grayish spots. Its larval foodplants consist of lechuguilla and related hybrids. At least two subspecies of this giant-skipper occur in the Trans-Pecos: *chinatiensis* in the Chinati Mountains region and *mariae* in the remainder of our area.

Glossary

Antennae: a pair of sensory appendages located on the head.

Basal: the portion of the wing nearest the body.

Cell: the interior area of a wing, enclosed by veins.

Club: the thickened end of the antennae.

Costal fold: a narrow flap of scent scales, used for sexual attraction, located near the front edge of the forewing.

Costal vein: a vein located on the upper edges of the wings.

Dimorphic: having two distinct forms.

Discal bar: a line on the center of the wings.

Dorsal: pertaining to the upper side.

Eyespot: a round spot on the wing, often with a darker or lighter interior, against a lighter or darker background.

Foodplant: plant utilized for egg laying on which the larvae feed.

Forewing: the front wing.

Hindwing: the rear wing.

Hyaline spot: a glassy or transparent area of the wing.

Larva: the caterpillar; an immature butterfly.

Margin: the edge of the wing.

Palps: paired appendages used for sensing and for cleaning the proboscis.

Proboscis: a coiled tube that extends from the mouth, used for feeding.

Ray: a narrow band of color that contrasts with the background color.

Scales: the tiny shingle-like structures that provide color.

Stigma: a patch of specialized scales on the forewing that produce a scent, used to attract a mate.

Vein: a narrow tube in a wing that provides structural support for the wing.

Ventral: pertaining to the underside.

Other Field Guides

Glassberg, Jeffrey. 2001. *Butterflies through Binoculars, the West*. New York: Oxford University Press.

Neck, Raymond W., and Geyata Ajivsgi. 1996. *A Field Guide to Butterflies of Texas*. Houston: Gulf Publishing.

Opler, Paul A., and Amy Bartlett Wright. 1999. *A Field Guide to Western Butterflies*. Boston: Houghton Mifflin.

Pyle, Robert M. 1981. *The Audubon Society Field Guide to North American Butterflies*. New York: Chanticleer Press.

———. 1995. *The Audubon Society Handbook for Butterfly Watchers*. Boston: Houghton Mifflin.

Scott, James A. 1986. *The Butterflies of North America*. Stanford, CA: Stanford University Press.

References

Hatch, S. L., K. N. Gandhi, and L. E. Brown. 1990. *Checklist of the Vascular Plants of Texas*. College Station: Texas Agricultural Experimental Station.

North American Butterfly Association. 2001. *Checklist & English Names of North American Butterflies*. 2d ed. Morristown, NJ: North American Butterfly Association.

Butterfly Checklist—West Texas Parks and Preserves

This region includes Big Bend National Park, Big Bend Ranch State Park, Black Gap Wildlife Management Area, Davis Mountains Preserve (Texas Nature Conservancy), Davis Mountains State Park, and Guadalupe Mountains National Park.

Key to Distribution:
B = Southern Big Bend Area (principally)
D = Davis Mountains (principally)
G = Guadalupe Mountains (principally)
W = Widespread

Key to Status:
c = common
f = fairly common
u = uncommon
r = rare
s = stray
p = periodic colonist
h = hypothetical (needs additional documentation)
(all in the proper habitat and time of year)

SWALLOWTAILS—FAMILY PAPILIONIDAE

Swallowtails—Subfamily Papilioninae

___ Pipevine Swallowtail *(Battus philenor)*: W, c
___ Black Swallowtail *(Papilio polyxenes)*: W, f
___ Giant Swallowtail *(Papilio cresphontes)*: W, r
___ Ornythion Swallowtail *(Papilio ornythion)*: B&G, s
___ Eastern Tiger Swallowtail *(Papilio glaucus)*: B, s
___ Two-tailed Swallowtail *(Papilio multicaudata)*: W, c
___ Ruby-spotted Swallowtail *(Papilio anchisiades)*: B, s

WHITES AND SULPHURS—FAMILY PIERIDAE

Whites—Subfamily Pierinae

___ Mexican Dartwhite *(Castastica nimbice)*: B, p
___ Florida White *(Appias drusilla)*: B, s
___ Spring White *(Pontia sisymbrii)*: G, r
___ Checkered White *(Pontia protodice)*: W, c
___ Cabbage White *(Pieris rapae)*: W, r
___ Great Southern White *(Ascia monuste)*: B, s
___ Sara Orangetip *(Anthocharis sara)*: D, s

Sulphurs—Subfamily Coliadinae

___ Clouded Sulphur *(Colias philodice)*: W, f
___ Orange Sulphur *(Colias eurytheme)*: W, c
___ Southern Dogface *(Colias cesonia)*: W, c
___ White Angled-Sulphur *(Anteos clorinde)*: B&G, s
___ Yellow Angled-Sulphur *(Anteos maerula)*: B, s
___ Cloudless Sulphur *(Phoebis sennae)*: W, u
___ Orange-barred Sulphur *(Phoebis philea)*: B, s
___ Large Orange Sulphur *(Phoebis agarithe)*: W, u
___ Statira Sulphur *(Phoebis statira)*: D, s
___ Lyside Sulphur *(Kricogonia lyside)*: B, c; D&G, r
___ Boisduval's Yellow *(Eurema boisduvaliana)*: B, u; G, s
___ Mexican Yellow *(Eurema mexicana)*: W, f
___ Tailed Orange *(Eurema proterpia)*: B&D, s
___ Little Yellow *(Eurema lisa)*: B&D, u
___ Mimosa Yellow *(Eurema nise)*: B, s
___ Sleepy Orange *(Eurema nicippe)*: W, c
___ Dainty Sulphur *(Nathalis iole)*: W, c

GOSSAMER-WING—FAMILY LYCAENIDAE

Hairstreaks—Subfamily Theclinae

___ Colorado Hairstreak *(Hypaurotis crysalus)*: G, s
___ Great Purple Hairstreak *(Atlides halesus)*: W, u
___ Soapberry Hairstreak *(Phaeostrymon alcestis)*: W, r
___ Oak Hairstreak *(Satyrium favonius)*: G, u
___ Poling's Hairstreak *(Satyrium polingi)*: B&D, r
___ Brown Burro Hairstreak *(Fixsenia ?)*: B, s*

*not yet described (C. Durden, personal communication)

___ Amyntor Greenstreak *(Cyanophrys amyntor)*: B, s
___ Xami Hairstreak *(Callophrys xami)*: B&D, s
___ Sandia Hairstreak *(Callophrys mcfarlandi)*: B&D, u; G, c
___ Henry's Elfin *(Callophrys henrici)*: W, f
___ Thicket Hairstreak *(Callophrys spinetorum)*: G, f
___ Juniper Hairstreak *(Callophrys gryneus)*: B, r; D&G, f
___ Gray Hairstreak *(Strymon melinus)*: W, c
___ Red-lined Scrub-Hairstreak *(Strymon bebrycia)*: B, s
___ Mallow Scrub-Hairstreak *(Strymon istapa)*: B, s
___ Bromeliad Scrub-Hairstreak *(Strymon serapio)*: B, s
___ Leda Ministreak *(Ministrymon leda)*: W, r
___ Gray Ministreak *(Ministrymon azia)*: B, r
___ Dusky-blue Groundstreak *(Calycopis isobean)*: G, s
___ Arizona Hairstreak *(Erora quaderna)*: B, s

Blues—Subfamily Polyommatinae

___ Western Pygmy-Blue *(Brephidium exile)*: W, f
___ Cassius Blue *(Leptotes cassius)*: B, s
___ Marine Blue *(Leptotes marina)*: W, c
___ Cyna Blue *(Zizula cyna)*: W, r
___ Ceraunus Blue *(Hemiargus ceraunus)*: B, u; D&G, s
___ Reakirt's Blue *(Hemiargus isola)*: W, c
___ Eastern Tailed-Blue *(Everes comyntas)*: B&D, s
___ Western Tailed-Blue *(Everes amyntula)*: G, r
___ Spring Azure *(Celastrina ladon)*: W, f
___ Rita Blue *(Euphilotes rita)*: B, r
___ Melissa Blue *(Lycaeides melissa)*: D, r; G, f
___ Acmon Blue *(Plebejus acmon)*: B&D, u; G, f
___ Lupine Blue *(Plebejus lupini)*: D, s

METALMARKS—FAMILY RIODINIDAE

___ Fatal Metalmark *(Calephelis nemesis)*: W, f
___ Rounded Metalmark *(Calephelis perditalis)*: B, s
___ Rawson's Metalmark *(Calephelis rawsoni)*: B&D, u
___ Zela Metalmark *(Emesis zela)*: B, s
___ Mormon Metalmark *(Apodemia mormo)*: W, f
___ Hepburn's Metalmark *(Apodemia hepburni)*: B, r
___ Palmer's Metalmark *(Apodemia palmeri)*: W, f
___ Chisos Metalmark *(Apodemia nais chisosensis)*: B, f

BRUSHFOOTED BUTTERFLIES—FAMILY NYMPHALIDAE

Snouts—Subfamily Libytheinae

___ American Snout *(Libytheana carinenta)*: W, c

Heliconians and Fritillaries—Subfamily Heliconiinae

___ Gulf Fritillary *(Agraulis vanillae)*: W, u
___ Mexican Silverspot *(Dione moneta)*: B, s
___ Isabella's Heliconian *(Eueides isabella)*: B, s
___ Zebra Heliconian *(Heliconius charithonia)*: B, p; G, s
___ Variegated Fritillary *(Euptoieta claudia)*: W, c
___ Mexican Fritillary *(Euptoieta hegesia)*: B, s

True Brushfoots—Subfamily Nymphalinae

___ Dotted Checkerspot *(Polydryas minuta)*: W, r
___ Theona Checkerspot *(Thessalia theona)*: W, u
___ Chinati Checkerspot *(Thessalia chinatiensis)*: B&D, f; G, r
___ Fulvia Checkerspot *(Thessalia fulvia)*: B&D, u; G, f
___ Bordered Patch *(Chlosyne lacinia)*: W, c
___ Definite Patch *(Chlosyne definita)*: B&D, s; G, u
___ Crimson Patch *(Chlosyne janais)*: B&G, s
___ Tiny Checkerspot *(Dymasia dymas)*: W, f
___ Elada Checkerspot *(Texola elada)*: W, f
___ Texan Crescent *(Phyciodes texana)*: W, c
___ Pale-banded Crescent *(Phyciodes tulcis)*: B, s
___ Vesta Crescent *(Phyciodes vesta)*: W, f
___ Phaon Crescent *(Phyciodes phaon)*: W, f
___ Pearl Crescent *(Phyciodes tharos)*: W, f
___ Painted Crescent *(Phyciodes picta)*: B&D, r; G, f
___ Mylitta Crescent *(Phyciodes mylitta)*: G, s
___ Question Mark *(Polygonia interrogationis)*: W, u
___ Satyr Comma *(Polygonia satyrus)*: G, s
___ Mourning Cloak *(Nymphalis antiopa)*: W, f
___ American Lady *(Vanessa virginiensis)*: W, c
___ Painted Lady *(Vanessa cardui)*: B&D, f; G, c
___ West Coast Lady *(Vanessa annabella)*: W, r
___ Red Admiral *(Vanessa atalanta)*: B&D, c; G, u

___ Common Buckeye *(Junonia coenia)*: W, f
___ Tropical Buckeye *(Junonia genoveva)*: W, r
___ White Peacock *(Anartia jatrophae)*: D&G, s
___ Malachite *(Siproeta stelenes)*: B&G, s

Admirals and Relatives—Subfamily Limenitidinae

___ Red-spotted Admiral *(Limenitis arthemis)*: W, f
___ Viceroy *(Limenitis archippus)*: W, u
___ California Sister *(Adelpha bredowii)*: W, c
___ Mexican Bluewing *(Myscelia ethusa)*: B, s
___ Mexican Eighty-eight *(Diaethria asteria)*: B, s
___ Common Mestra *(Mestra amymone)*: B&D, f; G, s
___ Red Rim *(Biblis hyperia)*: B, s
___ Tailed Cecropian *(Historis acheronta)*: B, s
___ Many-banded Daggerwing *(Marpesia chiron)*: B, s
___ Ruddy Daggerwing *(Marpesia petreus)*: B, s

Leafwings—Subfamily Charaxinae

___ Tropical Leafwing *(Anaea aidea)*: B, c; D&G, u
___ Goatweed Leafwing *(Anaea andria)*: W, u

Emperors—Subfamily Apaturinae

___ Hackberry Emperor *(Asterocampa celtis)*: W, f
___ Empress Leilia *(Asterocampa leilia)*: B&D, u
___ Tawny Emperor *(Asterocampa clyton)*: W, f

Satyrs—Subfamily Satyrinae

___ Canyonland Satyr *(Cyllopsis pertepida)*: B, r; D&G, c
___ Red Satyr *(Megisto rubricata)*: W, c
___ Common Wood-Nymph *(Cercyonis pegala)*: B, s
___ Mead's Wood-Nymph *(Cercyonis meadii)*: W, f

Monarchs—Subfamily Danainae

___ Monarch *(Danaus plexippus)*: W, f
___ Queen *(Danaus gilippus)*: W, c
___ Soldier *(Danaus eresimus)*: B, s
___ Tiger Mimic-Queen *(Lycorea cleobaea)*: B, s

SKIPPERS—FAMILY HESPERIIDAE

Firetips—Subfamily Pyrrhopyginae

___ Dull Firetip *(Pyrrhopyge araxes)*: B, s

Spread-wing Skippers—Subfamily Pyrginae

___ Silver-spotted Skipper *(Epargyreus clarus)*: W, s
___ Hammock Skipper *(Polygonus leo)*: D&G, s
___ White-striped Longtail *(Chiodes catillus)*: B, s
___ Zilpa Longtail *(Chiodes zilpa)*: B&D, s
___ Short-tailed Skipper *(Zestusa dorus)*: D&G, s
___ White-crescent Longtail *(Codatractus alcaeus)*: D, s
___ Arizona Skipper *(Codatractus arizonensis)*: B&D, s
___ Long-tailed Skipper *(Urbanus proteus)*: D, s
___ Dorantes Longtail *(Urbanus dorantes)*: B, s
___ Golden Banded-Skipper *(Autochton cellus)*: B&D, f
___ Chisos Banded-Skipper *(Autochton cincta)*: B, s
___ Desert Cloudywing *(Achalarus casica)*: B&D, u
___ Northern Cloudywing *(Thorybes pylades)*: W, c
___ Drusius Cloudywing *(Thorybes drusius)*: B&D, u
___ Acacia Skipper *(Cogia hippalus)*: B, f; D, u
___ Golden-headed Scallopwing *(Staphylus ceos)*: B&D, f; G, r
___ Mazans Scallopwing *(Staphylus mazans)*: B, s
___ Texas Powdered-Skipper *(Systasea pulverulenta)*: B&D, f; G, r
___ Arizona Powdered-Skipper *(Systasea zampa)*: B&D, r; G, u
___ Sickle-winged Skipper *(Achlyodes thraso)*: B, s
___ Hermit Skipper *(Grais stigmatica)*: B&D, s
___ Sleepy Duskywing *(Erynnis brizo)*: B&D, u; G, c
___ Juvenal's Duskywing *(Erynnis juvenalis)*: B&D, u
___ Rocky Mountain Duskywing *(Erynnis telemachus)*: W, f
___ Meridian Duskywing *(Erynnis meridianus)*: B&D, u; G, c
___ Scudder's Duskywing *(Erynnis scudderi)*: B, r
___ Mournful Duskywing *(Erynnis tristis)*: W, u
___ Funereal Duskywing *(Erynnis funeralis)*: W, f
___ Small Checkered-Skipper *(Pyrgus scriptura)*: W, r
___ Common Checkered-Skipper *(Pyrgus communis)*: W, c
___ Tropical Checkered-Skipper *(Pyrgus oileus)*: D, s
___ Desert Checkered-Skipper *(Pyrgus philetas)*: B&D, f
___ Erichson's White-Skipper *(Heliopetes domicella)*: B&D, u; G, s

___ Common Streaky-Skipper *(Celotes nessus)*: B, f; D&G, u
___ Scarce Streaky-Skipper *(Celotes limpia)*: B&D, u
___ Common Sootywing *(Pholisora catullus)*: W, f
___ Mexican Sootywing *(Pholisora mejicana)*: B&D, r
___ Saltbush Sootywing *(Hesperopsis alpheus)*: B&D, u

Skipperlings—Subfamily Heteropterinae

___ Russet Skipperling *(Piruna pirus)*: D, s
___ Four-spotted Skipperling *(Piruna polingi)*: D, p
___ Chisos Skipperling *(Piruna haferniki)*: B, r

Grass-Skippers—Subfamily Hesperiinae

___ Julia's Skipper *(Nastra julia)*: B, u
___ Clouded Skipper *(Lerema accius)*: B&D, u; G, s
___ Tropical Least Skipper *(Ancyloxphya arene)*: W, r
___ Edward's Skipperling *(Oarisima edwardsii)*: B, r; D&G, f
___ Orange Skipperling *(Copaeodes aurantiacus)*: W, c
___ Southern Skipperling *(Copaeodes minimus)*: B, f
___ Sunrise Skipper *(Adopaeoides prittwitzi)*: D, u
___ Fiery Skipper *(Hylephila phyleus)*: W, u
___ Morrison's Skipper *(Stinga morrisoni)*: B, r; D, f; G, u
___ Uncas Skipper *(Hesperia uncas)*: B, s; D&G, u
___ Apache Skipper *(Hesperia woodgatei)*: W, r
___ Pahaska Skipper *(Hesperia pahaska)*: B&D, r; G, c
___ Green Skipper *(Hesperia virdis)*: D&G, u
___ Rhesus Skipper *(Polites rhesus)*: D&G, u
___ Carus Skipper *(Polites carus)*: D&G, r
___ Whirlabout *(Polites vibex)*: D, s
___ Sachem *(Atalopedes campestris)*: W, f
___ Delaware Skipper *(Anatrytone logan)*: B, r; D&G, u
___ Taxiles Skipper *(Poanes taxiles)*: B, r
___ Umber Skipper *(Poanes melane)*: B&D, u
___ Dun Skipper *(Euphyes vestris)*: W, r
___ Viereck's Skipper *(Atrytonopis vierecki)*: W, f
___ White-barred Skipper *(Atrytonopis pittacus)*: W, u
___ Python Skipper *(Atrytonopis python)*: W, u
___ Sheep Skipper *(Atrytonopis edwardsii)*: B, f; D&G, r
___ Simius Roadside-Skipper *(Amblyscirtes simius)*: W, u
___ Cassus Roadside-Skipper *(Amblyscirtes cassus)*: B, r; D, f

___ Bronze Roadside-Skipper *(Amblyscirtes aenus)*: W, f
___ Oslar's Roadside-Skipper *(Amblyscirtes oslari)*: W, u
___ Texas Roadside-Skipper *(Amblyscirtes texanae)*: W, u
___ Slaty Roadside-Skipper *(Amblyscirtes nereus)*: B&D, u
___ Nysa Roadside-Skipper *(Amblyscirtes nysa)*: W, u
___ Dotted Roadside-Skipper *(Amblyscirtes eos)*: W, u
___ Celia's Roadside-Skipper *(Amblyscirtes celia)*: B&D, r
___ Orange-headed Roadside-Skipper *(Amblyscirtes phylace)*: B, r; D&G, f
___ Eufala Skipper *(Lerodea eufala)*: B&D, u; G, s
___ Ocola Skipper *(Panoquina ocola)*: B, s

Giant-Skippers—Subfamily Megathyminae

___ Orange Giant-Skipper *(Agathymus neumoegeni)*: W, u
___ Chisos Giant-Skipper *(Agathymus neumoegeni chisosensis)*: B, r
___ Mary's Giant-Skipper *(Agathymus mariae)*: W, r
___ Yucca Giant-Skipper *(Megathymus yuccae)*: W, r
___ Ursine Giant-Skipper *(Megathymus ursus)*: W, r

Foodplants

(names from Hatch, Gandhi, and Brown, *Checklist of the Vascular Plants of Texas*)

Index

F

G

H

J

K

L

M

N

O

P

Q